McGraw Hill Glencoe

New York, New York   Columbus, Ohio   Chicago, Illinois   Peoria, Illinois   Woodland Hills, California

Also Available from McGraw-Hill Professional Development

Glencoe/McGraw-Hill and McGraw-Hill Professional Development offer a suite of professional development products, designed to enrich content knowledge and develop a greater mastery of teaching strategies and techniques in mathematics instruction. These programs provide professional learning opportunities in various formats, which are described below.

The *Glencoe Educational Strategies in Mathematics Grades 6–12 Online Series* consists of five modules, including Assessment in Mathematics, Differentiated Instruction, English Learners, Literacy Strategies, and Standards-Based Instruction. Each module provides an online experience in which participants watch and listen to video clips of classroom lessons and interviews with a guest educator. Participants complete interactive activities that focus on ways to model the instructional strategies addressed by the classroom lesson. Modules may be used individually or combined for custom courses. University credit is available. For more information on the *Glencoe Educational Strategies in Mathematics Grades 6–12 Series,* visit math.glencoe.com or contact Glencoe/McGraw-Hill at 800-334-7344 or by e-mail at customer.service@mcgraw-hill.com.

The *Glencoe Mathematics Professional Development Online Secondary Series* includes five Algebra 1 units, three Geometry Units, and two Algebra 2 units. Each unit provides an online experience in which program users watch and listen to video clips of a classroom lesson, interviews with a guest educator, and an interview with the teacher of the video lesson. Users complete two interactive mathematics activities that focus on ways to model the content addressed by the classroom lesson, and they complete several other activities that explore strategies used during the classroom lesson but which have application across the spectrum of secondary mathematics. Units may be used individually or combined for custom courses. University credit is available. For more information on the *Glencoe Mathematics Professional Development Online Series,* visit math.glencoe.com or contact Glencoe/McGraw-Hill at 800-334-7344 or by e-mail at customer.service@mcgraw-hill.com.

The *Glencoe Mathematics Professional Development Online Pre-Algebra Series* includes five pre-Algebra units. Each unit provides an online experience in which program users watch and listen to video clips of a classroom lesson, interviews with a guest educator, and an interview with the teacher of the video lesson. Users complete two to four interactive mathematics activities that focus on ways to model the content addressed by the classroom lesson, and they complete several other activities that explore strategies used during the classroom lesson but which have application across the spectrum of pre-Algebra mathematics. Units may be used individually or combined for custom courses. University credit is available. For more information on the *Glencoe Mathematics Professional Development Online Series,* visit math.glencoe.com or contact Glencoe/McGraw-Hill at 800-334-7344 or by e-mail at customer.service@mcgraw-hill.com.

Mc Graw Hill Glencoe

The McGraw-Hill Companies

Send all inquiries to:
Glencoe/McGraw-Hill
8787 Orion Place
Columbus, OH 43240-4027

ISBN 0-07-869708-5 *Educational Strategies Video Workshop Facilitator's Guide*

Printed in the United States of America.

4 5 6 7 8 9 10 045 14 13 12 11 10 09 08 07 06

Table of Contents

Assessment in Mathematics

Components of Good Classroom Assessment

Directions: Listen to Lois Mosley's comments on the components of good classroom assessment. Record your notes regarding her comments about each type of assessment. Record your reasons for using each type in your instruction.

Type of Assessment	Lois Mosley's Comments	My Reasons for Using
Informal		
Formative		
Summative		

Think and Discuss: Discuss the three types of assessment that Lois Mosley describes in her assessment overview. Which type of assessment do you tend to use most often? Describe some of your reasons for using each type of assessment. Share any goals you have for improving your assessment practices.

Planning Assessment

Directions: Watch the *Department Planning Meeting* video segment. Record your observations of various activities and interactions, which took place during the meeting, that you believe were effective and will positively impact participants' instruction and assessment.

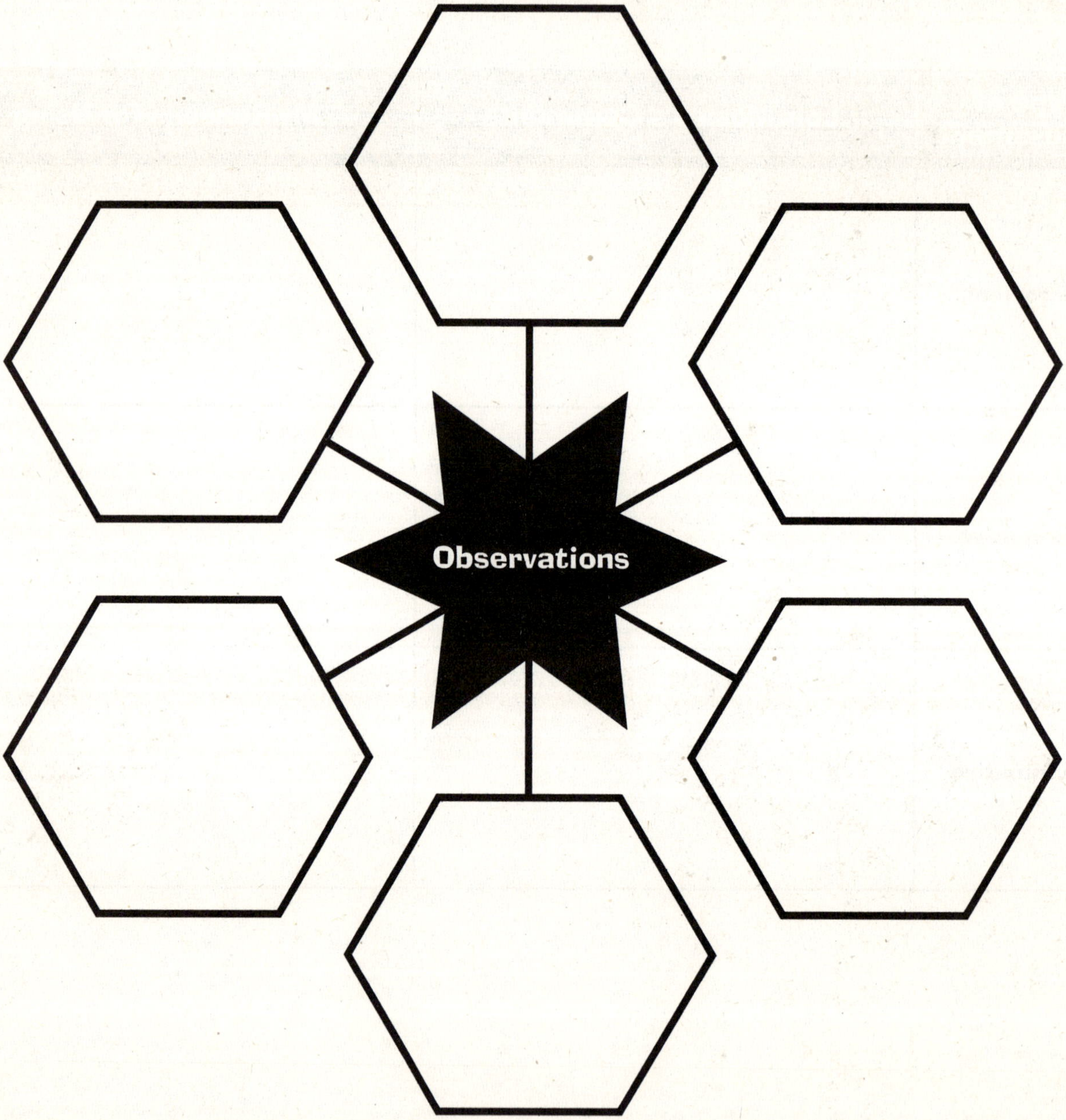

Think and Discuss: Discuss the department meeting and the process of planning assessment the team members completed. Describe the activities and interactions you observed during the meeting that you thought were effective. Describe any experiences you have had with collaborative assessment planning.

Designing Assessment

Directions: Consider the activities and discussion that the department members completed during their assessment planning meeting. You will create a similar assessment plan for a concept you will teach. Select a concept or standard that you will assess, and create an assessment item and the rubric you will use to evaluate your students' work.

Concept or Standard

Assessment Item

Evaluation Rubric			

Think and Discuss: Discuss the process of designing assessment. Describe some of the difficulties and challenges of planning assessment. Describe how carefully planned assessment can impact student achievement in a positive way. Share your assessment item and evaluation rubric.

Analyzing Results

Directions: In the segment, *Test Results and Analysis*, you saw the department use students' results to plan future assessment and instruction. Using the assessment plan you created, record the desired outcomes you hope to achieve when you analyze your assessment results.

Desired Outcomes

Identify student successes and needs.

Uncover any surprises regarding student understanding and performance.

Use the results to guide instruction.

Think and Discuss: Discuss the desired outcomes that you hope to achieve through your analysis of the assessment results. What do you hope to learn from what your students are able to demonstrate? How will your assessment help you uncover student misconceptions and gaps in understanding? What surprises might you find as you analyze the results? How will you use the results to inform your instruction?

Informal Assessment

Directions: Watch the *Informal Assessment* video segment. Observe the informal assessment strategies that Jason Willcoxon utilizes during his algebra lesson. Select a concept that you teach, and record two strategies you can use to informally assess your students' understanding of the concept.

Concept

Informal Assessment – Strategy One

Informal Assessment – Strategy Two

Think and Discuss: Discuss informal assessment strategies that you have found to be effective. Describe the informal assessment strategies you plan to implement for your selected concept.

Outline for an Assessment Planning Meeting

Directions: Discuss Lois Mosley's comments regarding the benefits of collaborative assessment planning. Reflect on the department planning meeting that you observed and the elements that made it successful. Create an outline for an assessment planning meeting. Record any additional notes you may have for the meeting.

Meeting Elements	Notes
Goal for Meeting	
Outline of Meeting Tasks	
Facilitator's Tasks	
Note-Taker's Tasks	

Additional Notes

Think and Discuss: Discuss the elements required to create successful collaborative planning meetings. Discuss the role and tasks of a facilitator, and the role and tasks of a note-taker. Describe your experiences as a participant in collaborative planning meetings.

Differentiated Instruction

Components and Goals

Directions: Watch the video segment, *Differentiated Instruction Overview* and listen as Nancy Frey discusses differentiated instruction and its components and goals. Think about an upcoming unit of study that you will teach to your students. Using the grid, record several examples of how each component of differentiated instruction can help you address each differentiated instruction goal.

Unit of Study or Topic: ______________________________

	Correct Misconceptions	Expand Background Knowledge	Extend Interests
Standards and Expectations			
Materials			
Grouping			

Think and Discuss: Discuss the basic components and goals of differentiated instruction as described by Nancy Frey. Describe some examples of student diversity that can be found among your students. Describe some varied levels of background knowledge and interests your students have.

Looking for Clues

Directions: As you view the two classroom lessons, look for evidence of differentiated instructional practices. How did Andrew Elizalde and Michael Cox differentiate their instruction? How did they anticipate the diversity of student needs and strengths? How did they create instruction that met the needs of a range of abilities and interests? Record the elements of a differentiated classroom you observed during the lessons.

Evidence of a Differentiated Lesson

Think and Discuss: What were the elements of these lessons that made them effective? What evidence of a differentiated classroom did you observe? What examples did you observe in which standards and expectations, materials, and groupings were differentiated? Did you observe any evidence in which the teachers corrected student misconceptions? Expanded background knowledge? Extended student interests?

Differentiated Materials and Activities

Directions: Andrew Elizalde's lesson on tangents illustrated several effective strategies for differentiated instruction. Mr. Elizalde carefully selected materials and activities that would help address the varied abilities, interests, and background knowledge of his students. Record several examples of the varied materials and activities he used during his lesson. How did the materials and activities benefit his students?

Example	How Used in Lesson	Benefits

Think and Discuss: Discuss the materials and activities that Mr. Elizalde selected for his lesson. Is there anything you would have done differently? If so, what other materials and activities would you have selected for a similar lesson?

Flexible Grouping

Directions: During his lesson on tangents, Andrew Elizalde successfully implemented several flexible grouping patterns. Record your observations of each type of grouping pattern Mr. Elizalde used during his lesson. Think of a similar lesson, and record several ideas for how you might implement each grouping pattern during that lesson.

Type of Grouping	Example Observed	Idea for Your Lesson
Teacher to Class		
Teacher to Individual or Small Group		
Student to Student		
Student to Class		

Think and Discuss: Discuss the various grouping arrangements that Andrew Elizalde implemented during his lesson. Were any particular groupings especially effective for his students and the lesson's subject matter? How did the grouping arrangements address the varied needs and preferences of Mr. Elizalde's students? Describe some of your ideas for how you could implement the same grouping patterns for a similar lesson.

Multiple Representations

Directions: After viewing the *Multiple Representations* video segment, think about how you can use multiple representations to model a new concept for an upcoming lesson or topic. In the table, record your ideas for using multiple representations in your lesson.

Concept to Model: ____

Multiple Representation Example	Idea for Your Lesson

Think and Discuss: Discuss the multiple representations you selected for your upcoming lesson or topic. Describe how each representation would address your students' various background knowledge, abilities, and interests.

Scaffolding Instruction

Directions: Nancy Frey discusses some of the effective elements of Andrew Elizalde and Michael Cox's lesson. Nancy Frey describes several critical elements of these lessons that enabled them to successfully differentiate their instruction. Think of an upcoming unit of study or topic you will teach, and list your ideas for how you can use scaffolding to help anticipate misconceptions and build conceptual understanding.

Upcoming Unit of Study or Topic: ____________________

Build Conceptual Understanding

Anticipate Misconceptions

Identify New Concept and/or Skill

Use of Scaffolding

Think and Discuss: Discuss the use of scaffolding as an instructional strategy for differentiating instruction. How can you use scaffolding to support and assist students in learning new concepts or procedures? Describe misconceptions to anticipate for your selected unit of study or topic. How can you use scaffolding to help build your students' conceptual understanding?

English Learners

Tapestry of Learners

Directions: Listen to Ruben Zepeda's comments regarding English learners. He discusses the benefits and challenges these students may bring to the classroom. Think of your students who are English learners, and list several examples of the unique experiences and perspectives these students bring to your classes. List several examples of the unique challenges these students may face in mathematics classes.

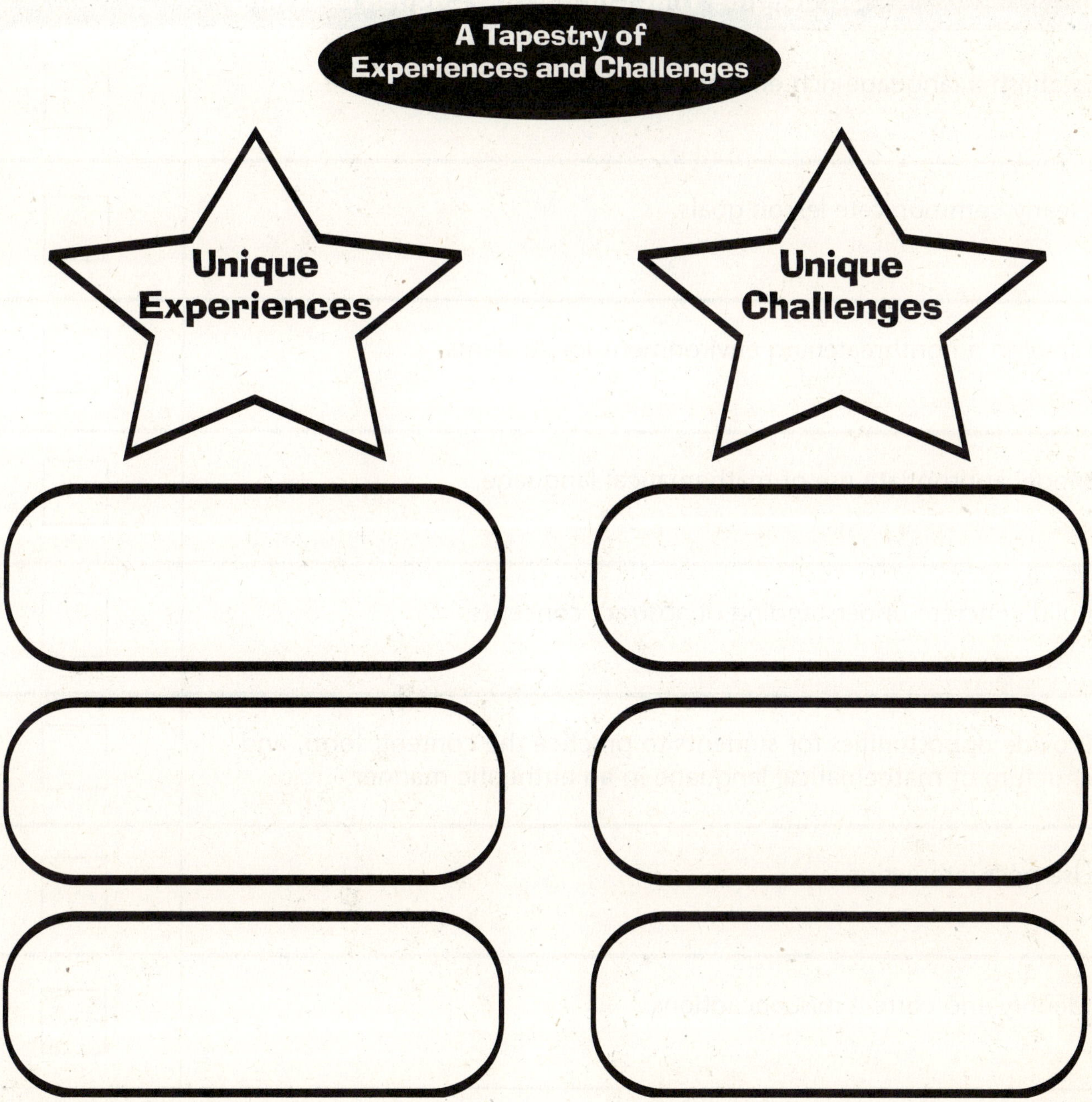

Think and Discuss: Discuss English learners in mathematics. Describe some of the unique characteristics of these learners and some of the unique challenges they may experience in your mathematics classes.

Strategies Checklist

Directions: As you watch the classroom lessons, use this checklist to look for strategies implemented by the two teachers to encourage English learning in mathematics classes. As you plan your mathematics lessons, you may also use this checklist to help you incorporate strategies that can help encourage English learning.

Strategies That Encourage English Learning	
Establish a language-rich environment.	☐
Clearly communicate lesson goals.	☐
Establish a nonthreatening environment for students.	☐
Model appropriate use of mathematical language.	☐
Build concrete understanding of abstract concepts.	☐
Provide opportunities for students to practice the content, form, and structure of mathematical language in an authentic manner.	☐
Use peer interaction.	☐
Identify and correct misconceptions.	☐

Think and Discuss: Discuss the two classroom lessons. Describe the strategies you observed that helped encourage English learning in mathematics classes. Discuss the lesson elements you thought were beneficial for English learners.

Communicating Effectively

Directions: After viewing the *Communicating Effectively* video segment, rate the effectiveness of Michael Cox's instruction in addressing each of the communication goals listed below. Reflect on your own instruction, and describe an example of how you have addressed each of these goals in your lessons.

Communication Goals	Rating of Effectiveness (1 = not effective, 2 = somewhat effective, 3 = effective, 4 = very effective)
Communicate expectations for learning.	○ 1 ○ 2 ○ 3 ○ 4
Provide frequent repetition of key vocabulary.	○ 1 ○ 2 ○ 3 ○ 4
Develop metacognitive skills to support learning.	○ 1 ○ 2 ○ 3 ○ 4

Examples from My Lessons:

____

____

____

____

____

____

Think and Discuss: Discuss Michael Cox's lesson and the strategies and techniques he implemented to facilitate communication. Discuss his effectiveness in meeting each of the communication goals. Describe an example of how you have addressed each of these goals in your own instruction.

Learning with Concrete Models

Directions: Andrew Elizalde and Michael Cox utilized concrete models in their mathematics lesson to help build conceptual understanding of abstract concepts. This strategy can be especially effective for English learners. Select three upcoming lesson topics or concepts for which you can use a hands-on activity to help students understand abstract concepts through concrete experiences.

Topic or Concept	Manipulative	Activity

Think and Discuss: Discuss Andrew Elizalde's and Michael Cox's use of concrete materials to help build conceptual understanding of abstract concepts. How can you transfer what you observed in their lessons to your own instruction? Describe a few activities in which you can use hands-on activities to help your students learn abstract concepts.

Creating a Dynamic Oral Culture

Directions: Andrew Elizalde's lesson on tangents demonstrated effective instructional strategies and techniques that can foster a language-rich learning environment. Many elements of his lesson can be helpful for English learners in mathematics classes. Record examples of each strategy that you observed in his lesson. Record an example of how you can implement the same strategy in your own instruction.

Strategy	Example from Andrew Elizalde's Geometry Lesson	How I Can Implement
Provide opportunities for students to ask and answer questions in a nonthreatening environment.		
Provide opportunities for students to express ideas verbally.		
Use pairs or small groups to encourage student interaction.		
Provide opportunities for students to help each other.		
Encourage communication through engaging activities.		

Think and Discuss: Discuss the strategies and techniques that Andrew Elizalde utilized to help foster a language-rich environment in his mathematics class. Discuss the elements of his instruction that encouraged communication. Describe how you can implement some of the same strategies in your lessons.

Lesson Analysis

Directions: Ruben Zepeda discusses Andrew Elizalde's and Michael Cox's lessons and the exemplary instructional practices they implemented to help address the needs of English learners. Listen to Mr. Zepeda's observations and record notes on the benefits of each instructional practice. After listening to his comments, record any additional thoughts you may have on the benefits of each instructional practice for English learners in mathematics.

Exemplary Instructional Practice	Ruben Zepeda's Comments	My Thoughts
Foster a language-rich environment.		
Provide multiple opportunities for students to learn and to demonstrate understanding.		
Build conceptual understanding of abstract concepts.		
Connect visual representations to mathematical concepts and vocabulary.		

Think and Discuss: How did Ruben Zepeda's comments add to your observations of each instructional practice? What new insights did his comments bring to your ideas on instructional practices that are beneficial for English learners?

Literacy in Mathematics

Students' Literacy in Mathematics

Directions: Listen to Douglas Fisher's comments regarding literacy in mathematics. Think of the four main components of literacy—reading, listening, writing, and speaking—and how these skills are exhibited in mathematics. Record the related proficiencies this skill implies. Then, think about your students' relative literacy levels, and rate their literacy levels in each of the four main components.

My Students' Literacy in Mathematics

Literacy Area	Proficiencies	My Students' Literacy Level
Reading	Example: read and comprehend a word problem	○ ○ ○ ○ ○ Low High
Listening		○ ○ ○ ○ ○ Low High
Writing		○ ○ ○ ○ ○ Low High
Speaking		○ ○ ○ ○ ○ Low High

Think and Discuss: Discuss literacy in mathematics and what this means to you. Describe examples of how students demonstrate literacy in mathematics. Describe ways your students communicate about mathematics in your classes. Discuss your students' strengths and weaknesses in the four main literacy areas of reading, listening, writing, and speaking. Describe one or two goals for helping your students improve their literacy in mathematics.

Outline for a Lesson

Directions: After viewing the two classroom lesson segments that emphasized literacy skills in mathematics, think of a similar lesson you could lead, then create an outline for this lesson. List activities to incorporate into your lesson that will foster your students' literacy skills in mathematics. You may use examples from the two classroom lessons or list other activity ideas.

Lesson Topic: ____________________

	Suggested Goals	Example Seen	Your Activity
Warm-Up Activity	■ Build vocabulary concepts ■ Apply background knowledge		
Whole-Group Activity	■ Encourage communication about mathematics		
Small-Group Activity	■ Share ideas ■ Explain reasoning ■ Create presentations		
Independent Practice or Extension Activity	■ Clarify and explain thinking ■ Strengthen reading skills		

Think and Discuss: Describe some of the elements in the two sample classroom lessons that helped strengthen student literacy in mathematics. Discuss elements of these lessons you thought were especially effective. Share your ideas for a similar lesson. Describe how the activities will help encourage your students to communicate about mathematics.

Building Vocabulary Skills

Directions: In his algebra lesson, Michael Cox had students complete an effective vocabulary-building activity. Select a topic you teach that involves important vocabulary terms, and create a similar vocabulary chart for your students to complete. Fill in the entire chart, then select one cell in each row to include on the version you will present to your students.

Term or Property	Definition	Example

Think and Discuss: Describe any strategies, techniques, and activities you have found successful in helping students learn and apply mathematical vocabulary. How do you address any misconceptions your students may have with vocabulary terms?

Reading Strategies

Directions: Michael Cox led a reading activity in his algebra class that helped build his students' literacy skills. He utilized the strategy of working backwards and modeled a metacognitive approach to reading in mathematics. In Douglas Fisher's discussion of literacy in mathematics, he described a variety of strategies to help students strengthen their reading skills in content areas. Think of your students who may need extra help with their reading skills in mathematics, then select from the strategies you have seen to create an intervention plan you can use with these students. You may include additional strategies that you have found to be effective.

Strategies to Strengthen Reading Skills in Mathematics

- Activate and apply background knowledge.
- Establish a purpose for reading.
- Scan text for information.
- Use text features.
- Take notes.
- Create a graphic organizer to record what was read.
- Write a summary of what was read.
- Work backwards.
- Use metacognitive strategies.
- ______________________________
- ______________________________
- ______________________________

Intervention Plan

____

____

____

____

Think and Discuss: Describe any areas of concern you may have regarding your students' reading abilities in mathematics. Discuss intervention strategies you can implement to help address these students' needs.

Writing in Mathematics

Directions: Michael Cox led an engaging writing activity in his algebra lesson. Think about some of the writing activities you have implemented in your own math classes. Describe two of these activities. List an example of what went well and what you would do differently if you were to repeat the activity with another class.

My Writing Activities

Activity	
What Worked Well	
What I Would Do Differently Next Time	

Activity	
What Worked Well	
What I Would Do Differently Next Time	

Think and Discuss: Describe the writing activities you implemented in your math classes. Describe what went well and what you would do differently the next time. Discuss how writing activities can be used to strengthen students' literacy in mathematics.

Instructional Strategies

Directions: Listen to Douglas Fisher's observations of Michael Cox's and Doug Roberts' algebra lessons. Discuss the various strategies and activities the two teachers implemented that emphasized literacy skills. Record a few ideas from these lessons that you can implement in future lessons to help build your students' literacy in mathematics. Use these ideas and the list of instructional strategies as a reminder to incorporate into future lessons those strategies that will help build your students' literacy in mathematics.

Strategies and Activities to Implement in Future Lessons

____

____

____

____

Instructional Strategies to Build Literacy in Mathematics

- ☐ Vary lesson resources to include textbooks, magazines, graphs, charts, and technology.
- ☐ Build new vocabulary on existing schema and vocabulary.
- ☐ Integrate writing, speaking, and listening opportunities.
- ☐ Create opportunities for students to take effective notes.
- ☐ Provide appropriate reviews of content and academic vocabulary.
- ☐ Model effective reading strategies.
- ☐ Analyze and generate a variety of written formats used in mathematics.
- ☐ Use quick-writes, journals, or learning logs for daily writing activities.

Think and Discuss: Discuss the strategies and activities implemented by Michael Cox and Doug Roberts in their algebra lessons. Discuss the ways these two teachers helped build their students' literacy in mathematics. Share a few of the ideas you observed in these lessons that you plan to use in future lessons. Discuss the list of strategies. Which of these strategies do you intend to implement in future lessons? Which of these strategies do you already incorporate into your lessons?

Standards-Based Instruction

Standards-Based Instruction Is...

Directions: Listen to Emily Schell's comments on standards-based instruction. Record two additional comments from Emily's discussion regarding the characteristics of standards-based instruction and the impact it can make on instruction. List two of your opinions regarding what standards-based instruction means to you.

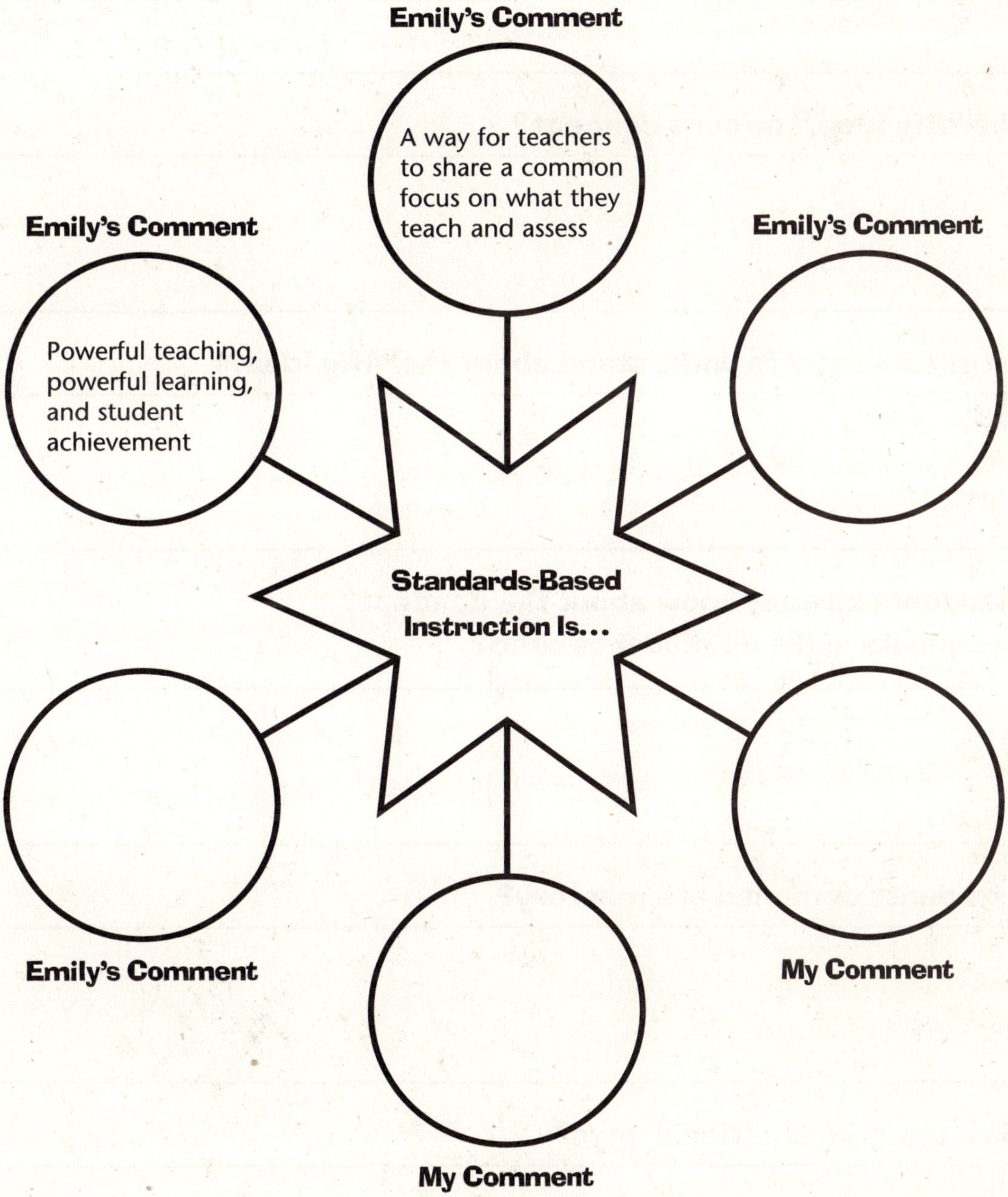

Think and Discuss: What does standards-based instruction mean to you? How has standards-based instruction changed the way you plan, teach, and assess? How well do you think your instructional practices address the standards for your grades and content areas?

Planning Standards-Based Instruction

Directions: Watch the *Department Planning Meeting* video segment. After observing the teachers completing an analysis of a standard, work in a small group to complete a similar activity.

Identify the standard(s).

What is the "Big Idea," or core concept?

What do students need to understand about the "Big Idea?"

What do students already know about the concept?
What prerequisite skills do students need?

How will students demonstrate mastery?

Pacing: How many instructional days?

Think and Discuss: Discuss the department meeting and the process of analyzing a standard that the team members completed. Describe any strategies or interactions you observed that you thought were effective. Share your standard analysis with the rest of the group.

Analyzing Standards

Directions: In the video, you saw the department analyze a standard to identify the essential concepts, content, and skills associated with a standard. Discuss the benefits of analyzing standards, then rank the benefits in order of importance for your own instruction. Discuss the ranking that you selected for the benefits.

Ranking of Importance

Benefit	Rank (1 to 5)
Understanding the "Big Idea"	☐
Identifying what students need to understand about the "Big Idea"	☐
Identifying prerequisite skills	☐
Identifying how students will demonstrate mastery	☐
Establishing a consistent curriculum among fellow teachers	☐

Think and Discuss: Discuss the benefits of analyzing standards. Share any additional benefits to analyzing standards. Describe an example of how you have experienced each of these benefits in your own planning and instruction.

Aligning Assessment to the Standards

Directions: Using the standard analysis you completed earlier, respond to each question regarding the assessment that will be used for the standard.

Questions to Ask

How will I know a standard has been met?

What assessment measure will provide evidence of proficiency?

Will multiple assessment measures be needed?

What is the criterion for demonstrating success?

Does the measure provide for more than one way to demonstrate success?

Think and Discuss: Discuss the organizing questions that can help you plan assessment that is aligned with standards. Share your responses to each question. Do you think standards-based assessment takes more planning time than traditional assessment? Discuss the impact this additional planning has on your instructional practices.

Communicating to Students

Directions: Watch the *Students and Standards* video segment. Observe the strategies Michael Cox utilizes to communicate standards to his students. Use this model, or another, to create a plan for communicating standards to your students. Record the materials you will use with your students to reference the standard and the procedure you will use to communicate the standard.

Strategies to Use

Standard

Materials

Procedure

Think and Discuss: Discuss the strategies that Michael Cox utilized to communicate the standards to his students. Describe your plan for communicating standards to your own students. Discuss the impact this awareness may have on your students. Do you think your students perform any differently when they are aware of the standards and what they are expected to demonstrate?

Value of Collaborative Planning

Directions: Emily Schell discusses the value of teachers collaborating on standards-based instructional planning. Listen to her observations and record notes on the value of teachers collaborating on planning standards-based instruction. After listening to her comments, record any additional thoughts that you may have on each listed benefit.

Benefits of Collaboration	Emily Schell's Comments	My Thoughts
Teachers share a common focus for instruction.		
Teachers share their expertise.		
Teachers participate in a focused conversation.		
Teachers take information back to the classroom to improve what they do.		

Think and Discuss: How did Emily Schell's comments add to your thoughts on the value of teachers collaborating on standards-based instructional planning? What new insights did her comments bring to your thoughts? Discuss some of the risks of collaborative planning. Do you think the rewards outweigh the risks?

Facilitator's Directions

Getting Started

Facitator's Direction

- **Purpose** The *Educational Strategies in Mathematics for Grades 6-12 Series Video Library Program* provides an opportunity for teachers to participate in a collegial discussion and an exchange of information and ideas that flows out of the content provided in the program video and handouts.

- **Materials** The program requires the use of both this Workshop Facilitator's Guide and the accompanying DVD, (name of DVD).

- **Workshop Activities** There are five videos and six professional development activities with associated handouts for each video. Together, the activities for a lesson unit take 1–2 hours to complete, including the time spent watching the videos. Activities can be completed by a group or as independent study.

- **Independent Study** Activities used in Independent Study should be completed in writing.

- **DVD Menu** There are five program videos included on the DVD. The DVD menu will allow you to navigate to the complete videos or to individual video segments that correspond to the workshop activities.

- **DVD Use** The videos for all activities are found on the DVD. The DVD is meant to be played on a television or on a computer with DVD movie capabilities. If you use either of these without a projector, you will need to ensure that all participants in the program are able to clearly see and hear the contents of the videos.

- **Facilitator's Role** Typically, a facilitator organizes the meetings, manages the setting (e.g., equipment, seating, and handouts), encourages participation on the part of attendees and, as needed, serves as a record keeper for attendance and participation. The program itself provides the content for participants to discuss.

- **Workshop Planning** You should read through lesson instructions and activities in advance to enable you to plan for appropriate seating, to decide if you want to use any materials other than those provided in the Workshop Facilitator's Guide (e.g., chart paper for shared results), and to identify the video associated with each activity.

- **Prior to Workshop** You should make copies of the activity handouts for all participants and preview the video referenced in the activity description. (Note: Some activities will suggest that a lesson video be played or replayed before the activity video is shown. In this event, you will need to allow extra time for the activity.)

- **During Workshop** Because professional interaction is a goal of the workshop, you should encourage participants to share their ideas and concerns about teaching. Each activity and group of questions is designed to promote this interaction with your help.

- **Facilitator Directions** Detailed directions for all activities are found in the pages that follow.

- **Ordering Information** 800-334-7344 or customer.service@mcgraw-hill.com.

Assessment in Mathematics

Page 3: Components of Good Classroom Assessment

- Have participants view the *Assessment Overview* video segment. Discuss with participants Lois Mosley's comments regarding the components of a good assessment program.
- Have participants record notes about each type of assessment as they listen to Lois Mosley's comments. Have participants record a few examples of why they utilize each type of assessment in their instruction.
- Have participants share their notes with the rest of the group.

Think and Discuss

- Discuss the three types of assessment that Lois Mosley describes in her assessment overview. Which type of assessment do you tend to use most often? Describe some of your reasons for using each type of assessment. Share any goals you have for improving your assessment practices.

Alternate Discussion Topic

- Have participants describe any high-stakes and/or statewide testing to which they may be accountable. Have participants share their strategies for using informal and formative assessments to help inform their instruction and to prepare students for high-stakes testing.

Page 4: Planning Assessment

- Have participants view the *Department Planning Meeting* video segment. Have participants share their observations of the activities and interactions that took place during this assessment planning meeting.
- Discuss with participants some meeting elements that were successful and that will have a positive impact on the participants' instruction and assessment.
- Have participants share their observations with the rest of the group.

Think and Discuss

- Discuss the department meeting and the process of planning assessment the team members completed. Describe the activities and interactions you observed during the meeting that you thought were effective. Describe any experiences you have had with collaborative assessment planning.

Alternate Discussion Topic

- Have participants discuss the contrast between collaborative assessment planning and assessment planning *completed individually*. Have participants describe the advantages and disadvantages to each approach.

Assessment in Mathematics

Page 5: Designing Assessment

- Have participants view the *Test Design and Alignment* video segment. Discuss the process completed by the meeting participants for planning their assessment. Explain to participants that they will work in small groups to complete a similar exercise.
- Have participants work in small groups to plan assessment for a concept or standard they teach. Have participants create a sample test item that they can use to assess their students' mastery of the concept.
- Have participants create a rubric they can use to evaluate the assessment item. Explain to participants that a rubric typically has four performance levels, ranging from low (no response given) to high (exceeding expectations). Have participants create four levels of responses that they may expect to see in their students' work. Remind participants that in the department planning meeting, the teachers created their rubric by determining what a typical complete response would contain and then working backwards through partial or incomplete responses.

Think and Discuss

- Discuss the process of designing assessment. Describe some of the difficulties and challenges of planning assessment. Describe how carefully planned assessment can have a positive impact on student achievement. Share your assessment item and evaluation rubric.

Alternate Discussion Topic

- Have participants share alternative techniques to rubrics for evaluating student performance on assessment items. Have participants share any resources they use for developing rubrics.

Page 6: Analyzing Results

- Have participants view the *Test Results and Analysis* video segment. Discuss with participants the importance of analyzing assessment results. Have participants share the experiences they have had with collaborative assessment analysis.
- Discuss with participants the three main desired-outcomes categories they can achieve through analysis of assessment results. Explain to participants that they will work in small groups to record goals they hope to achieve through analysis of test results.
- Have participants share their goals with the rest of the group.

Think and Discuss

- Discuss the desired outcomes you hope to achieve through your analysis of the assessment results. What do you hope to learn from what your students are able to demonstrate? How will your assessment help you uncover student misconceptions and gaps in understanding? What surprises might you find as you analyze the results? How will you use the results to inform your instruction?

Alternate Discussion Topic

- Discuss with participants any hesitancy they may have with collaborative assessment analysis. Explain that some participants in a collaborative assessment meeting may be nervous about exposing their students' performance levels. Have participants discuss the intimidation that some teachers may feel when asked to share their students' test results.

Assessment in Mathematics

Page 7: Informal Assessment

- Have participants view the *Informal Assessment* video segment. Discuss with participants the strategies that Jason Willcoxon utilized to informally assess his students during his algebra lesson.
- Discuss with participants various informal assessment strategies they have used in their own instruction. Discuss with participants the critical role that questioning plays in the informal assessment of students. Have participants share the questioning techniques with which they have had success.
- Have participants work in pairs or small groups to record two informal assessment strategies they can use during a lesson on a selected concept. Have participants share their ideas with the rest of the group.

Think and Discuss

- Discuss informal assessment strategies you have found to be effective. Describe the informal assessment strategies you plan to implement for your selected concept.

Alternate Discussion Topic

- Have participants share any experiences they may have had in which students' formative assessment results did not match the results of informal assessment. Have participants discuss some of the reasons why this may have occurred.

Page 8: Outline for an Assessment Planning Meeting

- Have participants view the *Value of Collaborative Assessment* video segment. Discuss the benefits of collaborative assessment planning that Lois Mosley describes. Discuss with participants the role and tasks of a facilitator during a collaborative assessment planning meeting, and the roles and tasks for a note-taker.
- Have participants work in pairs or small groups to record an outline for a successful assessment planning meeting.
- Have participants share their meeting outlines with the rest of the group.

Think and Discuss

- Discuss the elements required for successful collaborative planning meetings. Discuss the role and tasks of a facilitator, and the role and tasks of a note-taker. Describe your experiences as a participant in collaborative planning meetings.

Alternate Discussion Topic

- Have participants share any resources they use to accumulate assessment items.

Differentiated Instruction

Page 11: Differentiated Instruction: Components and Goals

- Have participants view the *Differentiated Instruction Overview* video segment. Discuss with participants Dr. Frey's comments about differentiated instruction and some of the components and goals of this instructional practice.
- Have participants work with a partner to complete the differentiated instruction grid. Have participants select an upcoming unit of study or topic. Have participants reflect on Nancy Frey's discussion brainstorm ideas for how they can use the three main components of differentiated instruction to help address three broad goals of this instructional practice.
- Have participants share their ideas with the rest of the group.

Think and Discuss

- Discuss the basic components and goals of differentiated instruction as described by Nancy Frey. Describe some examples of student diversity that can be found among your students. Describe some of the varied levels of background knowledge and interests your students have.

Alternate Discussion Topic

- If time permits, lead an extended discussion on the elements of differentiated instruction. You may want to expand on and extend from Dr. Frey's comments about how standards and expectations, materials, and groupings can be used to address students' varied learning needs, interests, and abilities. Discuss with participants how the essential questions, or "Big Ideas," should be identified for each unit of study, and how a unit of study's purpose and goals should be communicated to students.

Differentiated Instruction

Page 12: Looking for Clues

- Have participants view the classroom lesson video, *Classroom Examples*. Have participants look for various examples of differentiated instructional practices during Andrew Elizalde's lesson on tangents and Michael Cox's lesson on solving two-step equations.
- Have participants work with a partner to discuss and record examples of differentiated instruction they observed in the lessons. Have participants share their observations with the rest of the group.
- Tell participants they will have an opportunity to examine three instructional strategies from these lessons in more depth in the upcoming instructional strategies video clips.

Think and Discuss

- What were the elements that made these lessons effective? What evidence of a differentiated classroom did you observe? What examples did you observe in which Standards and Expectations, Materials, and Groupings were differentiated? Did you observe any evidence which showed that the teachers corrected student misconceptions? Expanded background knowledge? Extended student interests?

Alternate Discussion Topic

- Describe the planning and preparation that was required for this lesson and how these tasks contributed to the lesson's success. What were some of the tasks that Mr. Elizalde needed to complete prior to his lesson? What student needs did he consider prior to the lesson?

Page 13: Differentiated Materials and Activities

- Have participants view the first instructional strategy clip, *Differentiated Materials and Activities.*
- Have participants record examples of the various materials and activities Andrew Elizalde selected and implemented for his lesson on tangents. Have participants list an example of how the materials and activities were implemented. Have participants record a few ideas for how the materials and activities benefited students during the lesson.

Think and Discuss

- Discuss the materials and activities Mr. Elizalde selected for his lesson. Is there anything you would have done differently? What other materials and activities would you have selected for a similar lesson?

Alternate Discussion Topic

- If time permits, you may want to extend the discussion on materials selection and activities within a differentiated classroom. Tell participants that it can be helpful to use a few organizing questions to help clarify their selections and to ensure that, in the selection process, they are addressing the varied needs of their students. Discuss the following organizing questions with participants:
  - How can I make the content accessible to students?
  - What materials and models can I use to help explain the content?
  - What kinds of groupings are best used for the learning tasks planned?

Page 14: Flexible Grouping

- Have participants view the second instructional strategy clip, *Flexible Grouping*.
- Have participants notice the various grouping patterns used during Andrew Elizalde's lesson. Have participants record their observations of each grouping-pattern example.
- Have participants also record several ideas for how they can use the same grouping patterns for a similar lesson of their own.

Think and Discuss

- Discuss the various grouping arrangements that Andrew Elizalde implemented during his lesson. Were any particular groupings especially effective for his students and the lesson's subject matter? How did the grouping arrangements address the varied needs and preferences of his students? Describe some of your ideas for how you could implement the same grouping patterns for a similar lesson.

Alternate Discussion Topic

- Discuss with participants the classroom management strategies and techniques that should be considered when implementing flexible groupings. Have participants share some of their successful strategies for ensuring that students understand the behaviors expected of them when they are working in groups.

Strategy Information:

Flexible Grouping

- An important element of a differentiated classroom is the use of flexible grouping. Flexible grouping enables communication about concepts to occur on multiple levels. Flexible grouping provides the opportunity for students to learn in a variety of settings.

These settings include:

whole-group instruction—introduce content and model strategies

small-group work—heterogeneous groups provide for peer collaboration and scaffolding of learning

individual work—students work one-on-one with the teacher; encourages metacognition

- Varied groupings should occur frequently. Teachers may group students for a variety of reasons, including student interest, skill, materials available, language ability, reasoning abilities, and background knowledge.
- Flexible groupings also allow for communication about concepts to occur on multiple levels. The most frequently used flexible grouping patterns are:
  - *Teacher to Class*
    Teachers utilize time with the whole class for content introduction, task and procedure directions, and clarifications.
  - *Teacher to Individual or Small Group*
    Teachers can leverage small-group and individual time for clarifying and directing student efforts.
  - *Student to Student*
    Students gain much from engaging in tasks that require them to take ownership of material and negotiate meanings with their peers.
  - *Student to Class*
    Students gain valuable experience by presenting their work or opinions to the entire class.

Differentiated Instruction

Page 15: Multiple Representations

- Have participants view the third instructional strategy clip, *Multiple Representations*. Discuss with participants the use of multiple representations that Michael Cox used for his algebra lesson.
- If desired, you may also want to discuss with participants the various multiple representations Andrew Elizalde used during his tangents lesson.
- Have participants select an upcoming lesson or topic and record several ideas for the various multiple representations they can utilize for their lesson. Discuss with participants how they can use the various multiple representations to address their students' varied background knowledge, abilities, and interests.

Think and Discuss

- Discuss the multiple representations you selected for your upcoming lesson or topic. Describe how you can use each representation to address your students' varied background knowledge, abilities, and interests.

Alternate Discussion Topic

- Have participants share their ideas for an extension activity that Michael Cox could have assigned after his lesson. Have participants describe a lesson that could be conducted next. What are some examples of how Michael Cox could extend the learning in his lesson to new applications and activities?

Page 16: Scaffolding Instruction

- Have participants view Nancy Frey's analysis of the classroom lessons: *Nancy Frey's Observations*. Discuss with participants her comments about Mr. Elizalde's and Mr. Cox's lessons. Discuss with participants Dr. Frey's comments about the instructional practices and strategies the two teachers implemented that anticipated students' misconceptions, built conceptual understanding, and scaffolded instruction.
- Discuss with participants some of the elements of scaffolded instruction. (See "Strategy Information: Scaffolding," below.)
- Have participants work in pairs to brainstorm ideas for how they can implement scaffolded instruction in an upcoming lesson to help address student misconceptions and build conceptual understanding.
- Have participants share their ideas with the rest of the group.

Think and Discuss

- Discuss the use of scaffolding as an instructional strategy for differentiating instruction. How can scaffolding be used to provide support and assistance for students as they learn new concepts or procedures? Describe some of the misconceptions you should anticipate for your selected unit of study or topic. How can you use scaffolding to help build your students' conceptual understanding?

Alternate Discussion Topic

- Discuss with participants their ideas for how informal assessment could be utilized to help monitor student understanding during scaffolded instruction. Have participants share their successful informal assessment strategies and techniques.

Strategy Information: Scaffolding

- Scaffolding—or "temporary support"—is an instructional practice which provides support to students when they are learning new or difficult content or processes. The instructional support,

Page 16: Scaffolding Instruction (continued)

or assistance, is gradually removed as students show independence and mastery. Scaffolding begins slowly and builds from what students already know and know how to do. Slowly, new content and skills are introduced, modeled, and practiced. As students show understanding, teachers guide them to be more independent. As students show independence and mastery, the scaffolding, or support, is removed.

- Scaffolding is an important instructional strategy in a differentiated classroom because it helps students of various abilities be successful. Both teachers and students can provide support. Varied materials and activities can also be used to provide support.
- A few guidelines for effective scaffolding include:
  - Begin instruction with what students know, then help students achieve a measure of success quickly.
  - In small steps, slowly build understanding from students' prior knowledge.
  - Have students practice the new content and tasks, to apply and internalize the material.
  - Help students become independent when they have shown understanding of the new material.
- The following framework (Larkin, 1998) may be helpful as a general guide for scaffolding throughout a lesson:

  Teacher Does It—Teacher models new content or task.

  Class Does It—Teacher and students work together to perform the new task.

  Group Does It—Students work in pairs or small groups to practice the task.

  Individual Does It—The student completes independent practice in which he or she can demonstrate mastery.
- The following steps are frequently implemented during scaffolded instruction (Rosenshine and Meister, 1992):

1. New concept is presented and modeled.
   Building from student prior knowledge and making meaningful and interesting connections are important during this step.
2. Students complete guided practice.
   It is important during guided practice that difficulty is carefully regulated. Guided practice should start with easier tasks and gradually increase in difficulty. It is also important to anticipate student errors, misconceptions, and areas of difficulty, and to plan and prepare assistance and supplemental activities accordingly.
3. Varied contexts are provided for student practice.
   Student practice should involve teacher-led practice, as well as students working in pairs and in small groups.
4. Feedback is provided.
   Use feedback to help guide students toward independence. Feedback can be used to reinforce and model new concepts and procedures.
5. Student responsibility is increased.
   Gradually increase the complexity and difficulty of material and tasks. Gradually decrease support and assistance. Check for student mastery.
6. Students are assigned independent practice.
   Practice activities should be designed to help students apply newly learned content and processes to new applications.

Resources

- For additional information, teachers may want to view the ASCD videotape: How to Scaffold Instruction for Student Success (ASCD, 2002).

- Larkin, M. J. (2001). Providing support for student independence through scaffolded instruction. TEACHING Exceptional Children, 34(1), 30-34.
- Rosenshine, B. & Meister, C. (1992). The use of scaffolds for teaching higher-level cognitive strategies. Educational Leadership, 49(7), 26-33.

Page 19: Tapestry of Learners

- Have participants view the *English Learners Overview* video segment. Discuss with participants Ruben Zepeda's overview to English learners. Discuss with participants the notion of a "rich tapestry of experiences of perspectives" that Ruben describes regarding what English learners can bring to the classroom environment.
- Have participants describe some of the unique characteristics of their English learners. Have participants describe some unique challenges that these students may face in mathematics classes.
- Discuss with participants some of the factors that impact English learners and their academic success. Some of these factors include:
  - language proficiency
  - cultural background
  - academic experiences
- Discuss with participants some of the non-academic factors that may affect English learners. Discuss how English learners may experience higher levels of anxiety than native speakers. In additional to language and cultural differences, English learners may have concerns associated with:
  - physical appearance
  - using language appropriately
  - teacher expectations
- Discuss with participants that, because high levels of anxiety can negatively impact student learning, it is crucial for teachers of English learners to foster an atmosphere that minimizes the apprehension these students may feel. Discuss with participants how one way to accomplish this is by celebrating the unique experiences and perspectives many English learners can bring to your classes. Discuss also the importance of making sure English learners feel they are making positive contributions to the learning environment.

Think and Discuss

- Discuss English learners in mathematics. Describe some of the unique characteristics of these learners as well as unique challenges they may experience in your mathematics classes.

Alternate Discussion Topic

- Have participants share any successful intervention strategies they have implemented to help address any of the unique challenges English learners may have experienced in their mathematics classes.

English Learners

Page 20: Strategies Checklist

- Have participants view the *Classroom Examples* video segment.
- Have participants use the strategies list as a "checklist" as they observe the lessons. Have them note which strategies they observe during the lessons.
- Discuss with participants the various instructional strategies and activities the two teachers implemented to encourage English learning. Have participants describe the benefits of each observed instructional strategy for English learners.
- Explain to participants that they can use the list of strategies as a reference for their future lessons. Explain that this list, as well as the activities they observed in the two lessons, can provide a source of instructional ideas they can use to help encourage English learning in mathematics.

Think and Discuss

- Discuss the two classroom lessons. Describe the strategies you observed that helped encourage English learning in mathematics classes. Discuss the lesson elements you thought were beneficial for English learners.

Alternate Discussion Topic

- Discuss with participants any strategies they have used that enabled them to spend one-on-one time working with English learners in their mathematics classes.

Page 21: Communicating Effectively

- Have participants view the *Communicating Effectively* video segment.
- Discuss with participants how English learners need support to build fluency in English. Their ability to understand and use vocabulary is essential to understanding mathematics content. Students also need support in understanding the goals and expectations for lessons.
- Discuss with participants the communication strategies and techniques Michael Cox implemented during his algebra lesson.
- Have participants rate the effectiveness of Michael Cox in addressing each communication goal in his lesson.
- Have participants think about and record examples of how they have addressed each communication goal in their own instruction. Have participants share their examples with the rest of the group.

Think and Discuss

- Discuss Michael Cox's lesson and the strategies and techniques he implemented to facilitate communication. Discuss his effectiveness in meeting each of the communication goals. Describe an example of how you have addressed each of these goals in your own instruction.

Alternate Discussion Topic

- Discuss with participants examples of mathematical vocabulary that can be especially challenging for English learners. Discuss with participants how terms that have multiple meanings, such as mean and average, may present problems for these students. Have participants share other examples of similar terminology and describe any helpful strategies or activities they have used to help English learners with difficult mathematical vocabulary.

English Learners

Page 22: Learning with Concrete Models

- Have participants view the *Learning with Concrete Models* video segment. Discuss with participants the hands-on activities that Andrew Elizalde and Michael Cox implemented during their mathematics classes. Discuss with participants how these activities helped English learners build conceptual understanding for abstract concepts.
- Have participants work in pairs or small groups to brainstorm ideas for how they can utilize concrete models to help their English learners build understanding of abstract concepts. Have participants record three activity ideas.
- Have participants share their ideas with the rest of the group.

Think and Discuss

- Discuss Andrew Elizalde's and Michael Cox's use of concrete materials to help build conceptual understanding of abstract concepts. How can you transfer what you observed in their lessons to your own instruction? Describe a few activities in which you can utilize hands-on activities to help your students learn abstract concepts.

Alternate Discussion Topic

- Discuss with participants the planning, preparation, and classroom management techniques that are necessary for a successful hands-on activity.

Page 23: Creating a Dynamic Oral Culture

- Have participants view the *Creating a Dynamic Oral Culture* video segment. Discuss with participants the strategies and techniques that Andrew Elizalde used to foster a language-rich environment.
- Have participants record examples of each strategy observed in Mr. Elizalde's lesson. Have participants record an idea for how they can implement each strategy in their lessons.
- Have participants share their implementation ideas with the rest of the group.

Think and Discuss

- Discuss the strategies and techniques that Andrew Elizalde used to foster a language-rich environment in his mathematics class. Discuss the elements of his instruction that encouraged communication. Describe how you can implement some of the same strategies in your lessons.

Alternate Discussion Topic

- Discuss with participants the use of cooperative groups as an instructional strategy that can help encourage student interaction. Have participants describe the classroom management strategies and techniques they have used to help ensure successful cooperative group activities.

English Learners

Page 24: Lesson Analysis

- Have participants view the *Ruben Zepeda's Observations* video segment. As participants listen to Ruben Zepeda's comments, have them record notes from his discussion about the benefits of each instructional practice.
- Discuss with participants Mr. Zepeda's comments about the two lessons and the instructional strategies the teachers implemented.
- Have participants record any additional thoughts they have about each instructional practice.

Think and Discuss

- How did Ruben Zepeda's comments add to your observations of each instructional practice? What new insights did his comments bring to your ideas on instructional practices that are beneficial for English learners?

Alternate Discussion Topic

- Have participants discuss how informal assessment can be utilized to help monitor English learners' understanding of concepts and procedures. Have participants describe an example of how they have used informal assessment to identify English learner misconceptions or gaps in understanding.

Literacy in Mathematics

Page 27: Students' Literacy in Mathematics

- Have participants view the *Literacy in Mathematics Overview* video segment. Discuss with participants Douglas Fisher's overview of literacy in mathematics.
- Discuss with participants factors that contribute to student literacy, such as prior knowledge and the ability to integrate reading and language skills in the content areas.
- Discuss with participants several important skills that students need in order to demonstrate literacy in mathematics. These skills include the ability to:
  - understand content and academic vocabulary
  - use reading strategies to read and comprehend mathematical text
  - communicate effectively about mathematics
- Discuss with participants the four main components of literacy: reading, listening, writing, and speaking. Ask participants to think about, discuss, and record the types of abilities and implied proficiencies associated with each of these components in mathematics. Have participants think about their students' ability levels in each of these four strands and select a rating for each strand that corresponds to the general literacy level of their students.
- Have participants share their ratings with the rest of the group. Have participants discuss the general strengths and weaknesses of their students in the four literacy areas and identify several goals they would like to target to improve the literacy levels of their mathematics students.

Think and Discuss

- Discuss literacy in mathematics and what this means to you. Describe examples of how students demonstrated literacy in mathematics. Describe ways your students communicate about mathematics in your classes. Discuss your students' strengths and weaknesses in the four main literacy areas of reading, listening, writing, and speaking. Describe one or two goals for helping your students improve their literacy in mathematics.

Alternate Discussion Topic

- Discuss with participants the National Council of Teachers of Mathematics' Communication Standards. These standards focus on thinking and speaking mathematically and are found at every grade level and relate to every lesson topic.
- National Council of Teachers of Mathematics' Communication Standards:

  Instructional programs from prekindergarten through grade 12 should enable all students to:

  - organize and consolidate their mathematical thinking through communication
  - communicate their mathematical thinking coherently and clearly to peers, teachers, and others
  - analyze and evaluate the mathematical thinking and strategies of others
  - use the language of mathematics to express mathematical ideas precisely
- Review the standards with participants. Have participants describe instructional strategies and activities they have implemented to help address these standards. Discuss with participants the relationship between the Communication standards and literacy in mathematics.

  For more information, you may reference the NCTM Web site at: http://www.nctm.org/

Literacy in Mathematics

Page 28: Outline for a Lesson

- Have participants view the *Literacy Strategies Overview* video segment. Discuss and record examples of the instructional strategies and activities the two teachers implemented to help build their students' literacy levels.
- Have participants work with a partner to discuss and record ideas for a similar lesson they can lead with their students. Explain that they may use ideas they observed in the two sample classroom lessons and/or include other activity ideas.
- Have participants share their lesson ideas with the rest of the group.

Think and Discuss

- Describe some of the elements in the two sample classroom lessons that helped strengthen student literacy in mathematics. Discuss the elements of these lessons you thought were especially effective. Share your ideas for a similar lesson. Describe how the activities will help encourage your students to communicate about mathematics.

Alternate Discussion Topic

- Discuss with participants how the use of manipulatives in a mathematics lesson can help encourage students to communicate about mathematics. Have participants describe any activities they have implemented involving the use of manipulatives that sparked lively discussion among students.

Page 29: Building Vocabulary Skills

- Have participants view the *Vocabulary Strategies* video segment. Discuss with participants the vocabulary activity that Michael Cox had his students complete during his algebra lesson. Explain to participants that they will work in small groups to prepare a similar chart to use with their students.
- Have participants work in pairs or small groups to create their vocabulary chart. Explain that they will prepare the chart to use with their students. Participants will first complete an entire chart. After participants have completed their chart, they should select one cell in each row to include on the version they will present to their students. Have participants share their completed charts with the rest of the group.

Think and Discuss

- Describe any strategies, techniques, and activities that you have found successful in helping students learn and apply mathematical vocabulary. How do you address any misconceptions with vocabulary terms your students may have?

Alternate Discussion Topic

- Have participants share their ideas for assessing student understanding of vocabulary terms. Have participants describe any successful assessment strategies they have implemented with their students. Have participants discuss how they can use the vocabulary chart activity to assess student understanding of vocabulary terms.

Literacy in Mathematics

Page 30: Reading Strategies

- Have participants view the *Reading in Mathematics* video segment. Discuss with participants the reading strategies that Michael Cox emphasized during his algebra lesson.
- If time permits, you may want to have participants also view the *Literacy in Mathematics Overview* video segment in which Douglas Fisher describes strategies to help students strengthen their reading skills in mathematics.
- Have participants describe some of their students who may need extra help reading mathematical text. Have participants work in pairs or small groups to create intervention plans for these students. Have participants select strategies observed in Michael Cox's lesson and discussed by Douglas Fisher in his overview. Explain to participants that they may include additional strategies they have found successful in strengthening students' reading skills in mathematics.
- Have participants share their intervention plans with the rest of the group.

Think and Discuss

- Describe any areas of concern you may have with your students' reading abilities in mathematics. Discuss intervention strategies you can implement to help address these students' needs.

Alternate Discussion Topic

- Discuss with participants various note-taking strategies they have used with their students. Have participants share their note-taking-activity ideas.

Page 31: Writing in Mathematics

- Have participants view the *Writing to Learn* video segment.
- Have participants record two examples of writing activities they have implemented in their mathematics lessons. Have participants list an example of what went well and what they would do differently the next time they used the activity.
- Have participants share their activities and reflections with the rest of the group.

Think and Discuss

- Describe the writing activities you implemented in your mathematics classes. Describe what went well and what you would do differently the next time. Discuss how writing activities can be used to strengthen students' literacy in mathematics.

Alternate Discussion Topic

- Have participants complete an activity similar to the one Michael Cox's students completed in his classroom lesson. Have participants work in small groups to write word problems to match equations. Begin the activity by having each group write two equations. Have groups switch equations with another group and use the new equations to write their word problems. Have participants share their word problems. Discuss with participants how they might implement the same activity with their students.

Literacy in Mathematics

Page 32: Instructional Strategies

- Have participants view the *Douglas Fisher's Observations* video segment. Discuss with participants Douglas Fisher's comments about the two lessons and the instructional strategies implemented by the teachers.
- Have participants think about the two lessons and the various strategies and activities they observed during the lessons. Have participants record a few ideas from these lessons that they would like to incorporate into their future lessons.
- Explain to participants that they can use the list of strategies as a reference for their future lessons. Explain to participants that this list, as well as the lessons they have observed and discussed, can provide a source of instructional ideas they can use to help strengthen their students' literacy in mathematics.

Think and Discuss

- Discuss the strategies and activities implemented by Michael Cox and Doug Roberts in their algebra lessons. Discuss the ways these two teachers helped build their students' literacy in mathematics. Share a few ideas you observed in these lessons that you plan to utilize in future lessons. Discuss the list of strategies. Which of these strategies do you intend to implement in future lessons? Which of these strategies do you already incorporate into your lessons?

Alternate Discussion Topic

- Discuss with participants the use of cooperative groups as an instructional strategy that can help encourage communication about mathematics. Have participants describe the classroom management strategies and techniques they have used to help ensure successful cooperative group activities.

Standards-Based Instruction

Page 35: Standards-Based Instruction Is . . .

- Have participants view the *Standards-Based Instruction Overview* video segment. Discuss with participants Emily Schell's overview of standards-based instruction. Have participants record two additional comments from Emily's discussion, then have them list two of their own opinions regarding standards-based instruction. Have participants share their opinions.
- Discuss with participants the degree of familiarity they have with their state standards in mathematics. Have participants discuss how well they think their instructional practices address the standards for their grades and content areas. Have participants share any goals for improving instructional alignment with standards.

Strategy Information: Standards-Based Instruction

- If time permits, and participants desire it, present an overview to standards-based instruction. Discuss the following information with participants.
- Standards-based instruction is the practice of using standards of learning to guide learning and instruction.
- Teachers using this approach **identify**, **teach**, and **assess** the content and skills, which the standards specify, that students should know and be able to perform at each grade level.
- Standards provide **clear** and **uniform** guidelines for what students are expected to know and be able to do in content-area subjects at each grade level. Standards make clear the expectations for student learning to all educational partners, including students, teachers, parents, and administrators.
- Standards are written to establish expectations that are both attainable and measurable and that can apply to all, not simply a select few, students at a specific grade level.
- The teacher's role in standards-based instruction includes the following activities:
  - **Analyze and interpret** the standards in terms of their conceptual framework and the discreet facts and skills to be taught.
  - **Select** resources and teaching strategies that will aid all students in meeting the standard.
  - **Plan** assessment to determine whether students have achieved the standard

Think and Discuss

- What does standards-based instruction mean to you? How has standards-based instruction changed the way you plan, teach, and assess? How well do you think your instructional practices address the standards for your grades and content areas?

Alternate Discussion Topic

- In Emily Schell's overview, she makes the comment that standards-based instruction has "brought more professionalism" to teaching. Discuss her comment and have participants share their thoughts on this statement.

Page 36: Planning Standards-Based Instruction

- Have participants view the *Department Planning Meeting* video segment. Explain to participants that they will work in small groups to complete a similar planning exercise.

- Discuss with participants the basic steps involved in both analyzing standards and planning instruction.

  1. Identify content standards that will be taught in an upcoming unit.
  2. Use the following Guiding Questions to help organize your thinking and your goals for instruction:
     - What is the "Big Idea," the central concept, or the core idea of the standard?
     - What do students need to understand about this Big Idea?
     - What are the essential questions that will help students understand this idea?
     - What are the critical skills and content knowledge a student should demonstrate to meet this standard?
  3. Through assessing your students' prior knowledge, identify the content knowledge your students currently have about this content.
  4. Identify the prerequisite skills students will need before they can master the standard.
  5. Determine whether the standards can be broken into smaller learning segments.
  6. Determine whether the standards can be clustered with other similar standards that overlap.

- Have participants work in small groups to complete the standard-analyzing process and share their plans with the rest of the participants.

Think and Discuss

- Discuss the department meeting and the process of analyzing a standard that the team members completed. Describe any strategies or interactions, which you observed during the meeting, that you thought were effective. Share your standard analysis with the rest of the group.

Alternate Discussion Topic

- Discuss with participants the various approaches they use when planning instructional units. Have participants discuss the advantages and disadvantages of standards-based instructional planning compared to other approaches.

Standards-Based Instruction

Page 37: Analyzing Standards

- Have participants view the *Analyzing Standards* video segment. Discuss with participants the various benefits that can result when teachers analyze standards.
- Have participants rank the importance of each benefit in terms of their instruction. Have participants share their rankings and discuss their opinions regarding the importance of each benefit.

Think and Discuss

- Discuss the benefits of analyzing standards. Share any additional benefits to analyzing standards. Describe an example of how you have experienced each of these benefits in your own planning and instruction.

Alternate Discussion Topic

- Have participants discuss some of the disadvantages to analyzing standards. Have participants share ideas for how potential disadvantages can be minimized or eliminated.

Page 38: Aligning Assessment to the Standards

- Have participants view the *Pacing and Assessment* video segment. Discuss with participants the role of assessment in standards-based instruction.
- Standards-based assessment will measure each student's proficiency against the standard. To ensure success, standards-based assessment should:
  - be aligned to the specified content and skills stated in the standard
  - allow students to show evidence that they have met the standard
  - be performed regularly to drive instruction and ensure results for all students
- Remind participants that proficiency is:
  - defined as what each student needs to do to be successful
  - clearly linked to the standard
  - often determined by a scoring guide or rubric
  - defined in clear and precise language for all to understand
- Discuss with participants the organizing questions that can be used to help ensure successful standards-based assessment. Explain to participants that they now will work in their previously used planning groups to answer the assessment questions as these questions relate to the assessment participants will use for the standard they analyzed earlier.

Think and Discuss

- Discuss the organizing questions that can help you plan assessment that is aligned with standards. Share your responses to each question. Do you think standards-based assessment takes more planning time than traditional assessment? Discuss the impact this additional planning has on your instructional practices.

Alternate Discussion Topic

- Have participants discuss any special considerations they must make when assessing English learners or students with special needs. Have participants share their strategies for adapting assessment to meet the varied needs of their students.

Standards-Based Instruction

Page 39: Communicating to Students

- Have participants view the *Students and Standards* video segment. Discuss with participants the strategies that Michael Cox utilized to communicate the standards to his students.
- Have participants work in pairs or small groups to brainstorm ideas for communicating standards to their students. Have participants record a plan for communicating standards. Explain that participants should record an example of a standard they will teach to their students in the near future, the materials they will use (such as hand-outs, overhead transparencies, textbooks), and the procedure they will follow to discuss, explain, and illustrate the standard to their students.
- Discuss with participants the student's role in standards-based instruction. Students:
  - **see** the purpose of the standard and its impact on their learning
  - **understand** what they are expected to know, do, and learn to demonstrate proficiency in the standard
  - **acknowledge** that proficiency in the standard is expected for all students, not just the high performers

Think and Discuss

- Discuss the strategies that Michael Cox utilized to communicate the standards to his students. Describe your plan for communicating standards to your students. Discuss the impact this awareness may have on your students. Do you think your students perform any differently when they are aware of the standards and what they are expected to demonstrate?

Alternate Discussion Topic

- Discuss with participants the role students perform under standards-based instruction. Have participants describe the responsibilities and accountability students have when standards-based instruction is implemented. Have participants share their thoughts on the impact this role has on students and on the classroom environment.

Page 40: Value of Collaborative Planning

- Have participants view the *Value of Collaborative Planning* video segment. As participants listen to Emily Schell's comments, have them record notes from her discussion on the benefits of collaborative planning.
- Have participants record any additional thoughts they have on the benefits of collaborative planning.
- Discuss with participants some of the risks of collaborative planning. Have participants share any experiences they have had with collaborative planning. Have participants share whether or not they have experienced any of the risks.

Think and Discuss

- How did Emily Schell's comments add to your thoughts on the value of teachers collaborating on standards-based instructional planning? What new insights did her comments bring to your thoughts? Discuss some of the risks of collaborative planning. Do you think the rewards outweigh the risks?

Alternate Discussion Topic

- Have participants share any strategies or techniques, with which they are familiar, that can be used to help minimize any potential risks involved in collaborative planning.